Jennie Benedict

A choice collection of tested receipts

With a chapter on preparation of food for the sick

Jennie Benedict

A choice collection of tested receipts
With a chapter on preparation of food for the sick

ISBN/EAN: 9783337201463

Printed in Europe, USA, Canada, Australia, Japan

Cover: Foto ©berggeist007 / pixelio.de

More available books at **www.hansebooks.com**

A CHOICE COLLECTION

—OF—

TESTED RECEIPTS

WITH A CHAPTER ON

PREPARATION OF FOOD FOR
THE SICK

COMPILED BY

MISS JENNIE C. BENEDICT

LOUISVILLE
JOHN P. MORTON AND COMPANY
1897
K.

PREFACE.

In publishing this book the compiler has endeavored to give only a few choice, thoroughly tested receipts in every department of cooking.

Proceeding upon the maxim that "quality is more desirable than quantity," she has eliminated many receipts that are good, but not as practical, or else very similar to others given, and has chosen only those she has found to be the best and prepared in the best way.

The compiler has selected, as nearly as possible, those receipts that will be helpful and practical to the housekeeper and cook who desires to prepare either a simple home luncheon or a more elaborate dish for the dinner party ; thus attempting to answer to a certain degree the question, "What shall I have for the next meal, and how shall I prepare it ?"

Dainty dishes suitable for the invalid and trays for the convalescent patient are included, prepared in the most nutritious and yet palatable way.

She has endeavored to give each receipt in the clearest and simplest manner, so that even an amateur at cooking can use the book with ease, thus suiting the needs of all, from the beginner to the more experienced housekeeper.

In order to give all the advantage of her experience you will find on the back pages of the book the addresses of a few firms whose goods she has thoroughly tested and found to be the best in every way, and by the using of which she has obtained the best results : particularly the Daisy Flour, the Kentucky Refining Company's Nonpariel Oil, Fleischmann's Compressed Yeast, "Favorite" Cooking Stove for Natural Gas, and the Claudine Flour Sifter.

JENNIE C. BENEDICT.

INDEX.

(5)

ENTREES.

SAUCES.

SALADS.

SIX DAINTY MENUS.

CHAPTER ON MENUS.

RECEIPTS

In this book which can be used on the chafing dish :

Fried oysters.
Panned oysters.
Fricasseed oysters with
    mushrooms.
Little pigs in blankets.
Chicken with asparagus.

Any of the sauces.
Creamed chicken.
Creamed eggs.
Creamed calf's brains.
Creamed sweetbreads.
Parisienne potatoes.

COOKING

And What Can Be Done With It.

We can not deny that the great question of to-day, forever spoken by the hearts of the majority of women, is, "What am I fitted for, what can I do, and how can I do it?"

Possibly many, in the time this paper is being read, will ask of themselves anxiously, "Now has come the time I must do something, and what?"

There are two considerations that confront us, and prevent us from choosing wisely and profitably:

First—We are always striving after something that will not affect our social standing in the eyes of our neighbors.

Second—When considering a branch of work we invariably scan the completed specimen of some brother artist, and are awed by the immensity, forgetting that we are not expected to present a rounded act at first, that patience and perseverance are the secret of all successes.

When debating a field of work, never look at your occupation through your neighbor's eye, for nine chances out of ten she doesn't know as much about it as you.

Do not expect your work to dignify and sustain you until you have labored for your work and dignified it. Stand out alone, and look it fairly in the face, to learn if you are suited to accept and uplift it, and enable it to help you. Be willing to accept the beginning without reference to the end, and in this is constituted the important subject of our talk on cooking and what can be done with it.

Now, to show how small a beginning in cooking grew to a prosperous employment, because earnestly and enthusiastically entered into, I am going to tell you of the venture I know most about—my own.

(11)

When in September, 1893, I found myself face to face with the fact that I must fall into ranks with the great army of working women, my thoughts turned at once to cooking. Knowing I could make fruit cakes, I began by taking orders for them among my friends, and made them in my mother's kitchen.   In a few weeks work began to increase, and the family kitchen became too small a workshop for this occupation which had been placed in my hands.   My fruit cakes had gone forth to usher in a larger thing than I expected. I found to my surprise that a goodly number demanded of me, "Get you a kitchen and cook for us."

I was dismayed—I hadn't a cent of capital.   I could have our large yard to build my workshop on, so I picked up my courage and went to a carpenter, stating my case to him, telling him I would pay him for his work as soon as I earned the money, if he would build me a simple little kitchen with a pantry connected, putting up shelves for my "tools."   He may have discerned energy and courage, or he may have tasted my fruit cakes at some time.   Any way he built the kitchen, and just before Thanksgiving, in 1893, I started to work in my plain little shop with no conveniences or embellishments, these to be added as I felt able, and I was surprised to see how, little by little, the kitchen bloomed forth into pots and pans, all paid for by the occupation growing larger and larger.

I sent my circulars all over the city, stating that I would take orders from a cup of chocolate to a large reception.

The Principals of two of our largest schools asked me to supply lunches for the children.   So steadily the work grew, until now I have my cook and man, and have added one convenience after another, so that the kitchen is furnished with all necessary appliances.

Now I supply my own linen and china for receptions or dinners, and furnish trays for the sick.   Growing fond of my work, I have studied it as a science and an art, and last year taught classes in cooking at home and in other cities, which I continue this year.

Now, as the receipts, not only for fruit cakes but for all departments in cooking, have been liked, I shall attempt my cook hook, and by Thanksgiving the last round of the ladder of cooking will be climbed—I shall be able to stand behind a cook book, it's compiler. All this from fruit cakes well done.

Now I do not mean to prove by this that any one can immediately start a kitchen without a cent of capital and make it a success. Neither do I mean to imply that when the kitchen is started, that is the end ; but I do mean to show that the avenue of cooking, as an industry, is open to those earnestly looking for a means of livelihood, and who still have a womanly love for the homely art of cooking.

In the story of the kitchen I have also tried to show that though capital is not necessary, the constant bearing in mind of a high ideal is the motto, " Every thing attempted of the best, nothing attempted that can not be made of practical use, and nothing ever attempted which involves debts, the paying of which is an uncertainty," should be religiously adhered to.

Remember that nothing is ever gained by scrimping the means or materials of cooking in any way. Not a grand beginning and a gradual descent to the mediocre ; but a more modest attempt with an honest standard which is never lowered, is the secret of success in cooking.

Cooking is an art, because it is capable of unlimited growth, because it embraces the study of form and color, and most of all because it compels the perfect union of all things in its domain. The cook must know just when to withhold the grain of pepper or the sprinkle of salt to obtain the flavor that dubs him an artist. Nor is this sufficient, he must understand the person for whom the meal is prepared and plan the one for the other, so that they will agree as one, and in this subtle combining does cooking become an art. And now we have, as it were, only the picture— the frame has yet to come. This nectar and ambrosia

2

must be placed before our Lucullus in fit settings. We must know his characteristics—must there be cut glass and fine linen, or a Sevre service on which to serve the precious viands—must we encircle it in flowers, or leave it in severity ungarlanded ?

Cooking is a science, because in the preparing of foods we are brought in contact with the laws that govern the animal, vegetable, and mineral kingdoms. Chemistry is our constant companion. Nor can a cook stop with what is already finished. She must be an inventor. She must see in a cabbage or turnip such possibilities that will transform it from its original state to a dish fit for the gods.

You probably think I have placed cooking under a rather high standard in this "glorious age" when I say that cooking is a science, an art, and a philosophy. But stop a minute. Did you ever consider that food and the preparing of food are the means by which our great social engine is supplied with energy ?

Did you ever consider that food retarded or advanced the work of our body and mind, and therefore retarded or advanced national progress to a degree by no means small ?

The great sun gives its energy to the seed below ground, which, under his influence, become the vast expanses of wood, and by the aid of the same energy that raised it, the forest flourishes, falls, and decays, becoming great peat-bogs, and finally coal to burst into warmth and flame and energy again. Do we consider that power small which gave such strength for such a transformation ?

Shall we despise the power that builds us up, enables us to have strength to labor ?

We started with a little kitchen, we have ended in the banquet hall of peoples. Is it not true ?

It is not the branch of work alone that lifts us to a higher sphere,
For man may choose the humblest path to find the great is near;
God gives us all our part to do, and with us lies the right
To leave our task unbeautified, or mighty in His sight.

In this short time I have tried to show the use of cooking, the dignity of cooking, and the place of cooking in the
present day. I have said nothing of its noble past, for
noble it certainly is. Among its devotees we find the
greatest of the world—scientists, artists, philosophers, poets,
and potentates, all have found it a friend. Dumas was also
a cook, Balzac honored it, and some of our best loved mem-
ories of the English classics are found to be Charles Lamb's
"Dissertation upon Roast Pig," Dickens' delicious concoc-
tions in "Pickwick Papers," and Thackeray's inimitable
description of a French chef. Nor have I stepped far
enough back in the ages, to the time when the feast was
honored as a religious ceremony, or to the grand old Greeks
who reverenced that which nourished their bodies, and
whose women, from the matron to the youngest maid, knew
the art of cooking. Their feasts we all know were monu-
ments of magnificence, the legends of which have come
down to us of to-day. Yet we have found that like any
truly great object, cooking may own a very simple beginning, and opens readily its prospects to the one earnestly
seeking it.

We have found that cooking may be used as an honor-
able and practical means of livelihood for those who find that
now it has become a necessity for them to enter the fields
of labor. We find this is not its only use ; but that it may
also teach, and that from this revered and womanly occupation we learn, as well as in the more exalted fields of the
"new woman" sphere.

We have found that cooking possesses the dignity of an
art, a science, and of a philosophy, and for its place in this
age, it is one of the still unseen powers that uplifts and
enables our great peoples to progress.

GLOSSARY.

SAUTE.—Saute is to fry in as little fat as possible—frying is to immerse in hot fat.

MARINATE.—To make salads successfully, the meat, or celery, or nuts should be placed in a dish and covered with three parts oil and one part vinegar, and a little salt, which is to marinate for several hours. Then any of the dressing which is not absorbed should be drained off, the salad mixed as desired, and the regular dressing poured over it.

BROWN STOCK.—To make Brown Stock successfully, take a four-pound soup bone, remove some of the meat from the bone, and then place the bone in the soup kettle with three quarts of cold water and let it boil on the back of the stove. Take the soup vegetables with a little parsley and two cloves and the meat which you have reserved from the soup bone, chop all fine and saute until brown. Pour into the boiling kettle and let all boil together slowly five or six hours. Remove from the fire, strain through a fine sieve, let it cool, and skim off the grease. Put away in a cool place and use as desired.

WHITE STOCK.—Take the liquor in which chicken or veal has been boiled, remove the meat and season, boil for fifteen minutes with a stalk of celery, a slice of onion, two slices of carrot, and a bay leaf, and a little salt and pepper. Strain and use as White Stock.

DILUTED EGG.—Where egg and crumbs are to be used in frying, always dilute one egg with two tablespoonsful of water. This will prevent a hard crust from forming on any thing that is fried, and will make just a delicate brown.

CAKE.—To obtain the best results in making cake where milk and baking powder are to be used, stir into the milk the baking powder, and add to the cake the last thing, for

(16)

in many cases where the baking powder is put into the flour some of it is lost, and the cake is not as light as it should be.

CAKE.—In plum puddings, fruit cakes, mince meat, etc., where spices and liquor are used, I find it more desirable to let the spices stand in the liquor for an hour or more before putting into the other ingredients.

WHIPPED CREAM.—Remember that a pint of cream whipped is not a pint of whipped cream. Be careful to notice always whether the receipt calls for whipped cream or cream whipped.

RICE.—It is a most satisfactory way to soak rice in cold water for an hour or more before using.

BREAD.*

POTATO ROLLS.

1 cup flour.	1 cup potatoes (which have been
¾ cup of lard.	put through a potato-ricer.
1 cup of milk.	2 eggs, well beaten.
½ cup of sugar (scant).	1 teaspoonful of salt.

1 cake of Fleischmann's Compressed Yeast, dissolved in 2 cups of lukewarm water.

Mix thoroughly the lard, potatoes, sugar, and salt; add the eggs, then the milk, and then the yeast. Set to rise for two hours; make into a soft dough by adding about a quart of flour, and set to rise again. Make into rolls or loaf, butter the top, and set to rise again; bake in a quick oven.

PLAIN ROLLS.

1 pint of milk.	2 tablespoonsful of butter.
2 tablespoonsful of sugar.	1 teaspoonful of salt.
3 cups of flour for sponge.	¼ cup of lukewarm water.

¼ cake of Fleischmann's Compressed Yeast.

Scald the milk and pour it over the butter, sugar, and salt. When cold, add the yeast cake dissolved in the lukewarm water, then add the flour to make the sponge; beat well; let it rise until light. Then add enough flour to knead; knead well—very thoroughly—and set to rise. When light, cut it down, shape into rolls, let rise again, and bake in a quick oven.

---

*I find excellent results from using the Claudine sifter in bread and cake making.—J. C. B.

(19)

BEATEN BISCUIT.

1 quart of flour.	¼ cup of lard.
1 cup of cold water.	½ teaspoonful of salt.

Add two tablespoonsful of milk with the water (to make them brown nicely). Rub the lard well into the flour, and add the milky water until you have a stiff dough. Work through a biscuit machine or beat with an iron until the dough is smooth and light. Bake in a moderate oven.

CORN MUFFINS.

1 pint of meal.	½ pint of milk.
1 tablespoonful of lard.	2 eggs.
1 heaping teaspoon baking powder.	½ teaspoon of salt.

Beat the eggs separately until very light. Then add to the yelks the meal, baking powder, and salt sifted together. Then the lard melted, then the milk, and when just ready to pour into the hot buttered rings add the whites of eggs beaten to a stiff, dry froth.

WAFFLES.

2 cups of flour.	1 teaspoonful of baking powder (heaping).
½ teaspoonful of salt.	
2 eggs, well beaten.	1½ tablespoonsful melted butter.
1 cup milk.	

Mix the flour, baking powder, and salt, and sift; then add the well-beaten yelks of two eggs, to which has been added the milk, and stir into the dry mixture. Add the melted butter, then the whites of the two eggs, beaten to a stiff froth. Then have the waffle irons very hot and well greased—pouring off any extra grease, leaving only enough to keep batter from sticking.

SOUPS.

CONSOMME.

2 lbs. lean beef (from the round).	1 small chicken.
	1 small onion.
2 ounces lean ham.	¼ small carrot.
2 sprigs parsley.	2 stalks celery.
2 bay leaves.	2 eggs.
6 cloves.	A little celery salt.
½ lemon (juice of same).	

Wipe and cut the beef into small pieces ; cut the chicken as for fricasseed chicken. Cover with cold water and stand on the back of the stove where it will slowly heat. Simmer gently for four hours. Fry out a slice of bacon, add the ham cut in dice, the onion and carrot sliced, saute to a delicate brown in two tablespoonsful of butter ; then add this to the stock with the remainder of the vegetables (cutting the celery in pieces) and a little thyme. Let the soup simmer for another hour, strain and stand away to cool. When cold, carefully remove the fat from the surface ; put in a kettle over the fire, add the whites and shells of two eggs beaten lightly, two tablespoonsful of cold water, a little celery salt, and the juice of half a lemon. Let it boil for five minutes, take from the fire and skim carefully, and strain through a cloth. When ready to serve, heat again and season with salt and pepper to taste. The soup should be perfectly clear, but amber in color.

ST. GERMAIN.

1 can peas.	Water as much as there is
½ onion.	liquor in the can.
Sprig of parsley.	A blade of mace.
½ teaspoonful of sugar.	1 teaspoonful salt.
½ teaspoonful of pepper.	3 cupsful brown stock.
A bit of bay leaf.	

Drain and mash the peas, add the water, reserving one-half cup of the peas, putting the remainder into the stew pan with the onion, bay leaf, parsley, mace, sugar, salt, and pepper; simmer gently for half an hour, mash thoroughly, and add the hot brown stock. Let it come to the boiling point and rub through a sieve. Thicken with one tablespoonful of butter and one heaping tablespoonful of flour. Cook ten minutes and add the whole peas.—*Miss Farmer.*

QUICK BOUILLON.

1 tablespoonful butter.	½ small onion, sliced.
1 ½ lbs. finely chopped lean beef (round being best).	1 stalk celery.
	½ chicken (bones well broken).
4 cloves.	2 slices carrot.
1 bay leaf.	2 sprigs parsley.
1 ½ pints cold water.	1 egg (white and shell).

Melt the butter and add the onion, cook until the onion is thoroughly browned; then add the beef (that from the round being best) and chicken, celery, cloves, carrot, bay leaf, parsley, and cold water. Cover the saucepan, and stand on the back of the stove where the water will slowly heat. Let it come to boiling point, then simmer gently for two hours. Strain, return to saucepan and bring to a boil. Beat the white of one egg, add a half cup of cold water until thoroughly mixed, crush the shell and add to the egg and water, and then to the boiling bouillon ; boil four minutes, let it stand one minute to settle, and strain through cheese cloth wrung out of cold water.

BLACK BEAN.

1 pint black beans.	1 large soup bone.
½ teaspoon ground cin-	Several cloves.
namon.	2 onions, fried.
A little allspice.	A few slices of lemon.
2 tablespoonsful of butter.	Salt and pepper to taste.
½ tumbler of sherry wine.	

Soak the beans over night. Rinse well and boil thoroughly in the morning. When thoroughly cooked, mash through a colander and add it to soup stock made from the soup bone. Add the seasoning and onions and butter. When ready to serve, add three hard boiled eggs sliced.

PUREE OF ASPARAGUS.

1 quart white stock.	1 can asparagus.
1 pint cream.	1 level tablespoon of but-
1 heaping teaspoonful flour.	ter.

Put a little more than a quart of white stock (either chicken or veal broth) on the fire with the asparagus and let them boil hard for fifteen minutes, then strain, pressing all the substance from the asparagus—reserve the tips of asparagus to serve in puree. Thicken the strained stock with the butter and flour, and just before serving add the cream, salt, and pepper.

Celery, peas, etc., can be used in the same way.

FISH

MEATS

SAUCES ENTREES

SALADS

Desserts

CAKES

ICES

MISC.

TOMATO PUREE.

1 can tomatoes.	1 pint brown stock.
1 bay leaf.	1 sprig parsley.
1 stalk of celery.	1 teaspoonful of sugar.
1 tablespoonful of butter.	Several slices of onion.

Put the tomatoes into a sauce pan with the brown stock, bay leaf, parsley, celery, and sugar ; simmer thoroughly ; put the onion and butter into the saute pan, and when the onion is thoroughly done—but not brown—add a tablespoonful of flour, and put all with the tomatoes ; season with salt and pepper. Pass the whole through a fine sieve or strainer—heat again and serve.

OYSTER BISQUE.

1 quart of oysters.	4 cups of cream.
1 slice of onion.	2 stalks of celery.
2 blades of mace.	1 sprig of parsley.
A bit of bay leaf.	⅓ cup of butter.
⅓ cup of flour.	Salt and pepper to taste.

Scald the oysters and the liquor, separate them after heating to boiling point, strain liquor through cheese cloth, reheat, and thicken with the butter and flour. Scald the milk with the other ingredients mentioned, remove seasonings, and add the milk to the oyster liquor, and then add the oysters. Serve hot with whipped cream on top.—*Miss Farmer.*

FISH.

LOBSTER TIMBALS.

2 slices of stale bread.	1 pound halibut or cod.
1 egg, 1 yelk.	4 tablespoonsful of rich
½ tablespoon onion chopped	cream.
fine.	3 tablespoonsful of butter.
2 tablespoonsful of flour.	1 tablespoon sherry wine.
⅔ cup of lobster.	Salt and pepper to taste.

Soak two slices of stale bread in water until soft, squeeze until entirely free from water, cook with a teaspoonful of butter, beating to the consistency of india rubber, then cool ; put one pound of halibut or cod through a meat chopper, and then pound in a mortar. Add gradually one-third cup of bread, one egg, one yelk, and four tablespoonsful of rich cream; beat well. Butter timbal molds and spread the bread mixture on sides and bottom ; fill with the following lobster filling:

Saute one-half tablespoonful of onion, chopped very fine, in three tablespoonsful of butter ; add two tablespoonsful of flour, one-half cup of rich cream, yelks of two eggs, salt and pepper to taste. When this thickens, add a tablespoonful of sherry wine and two thirds of a cup of chopped lobster; pour out to cool, fill the center of the timbals, cover with the fish, and cook in a hot oven in a pan of hot water ; serve with lobster sauce. This proportion makes six timbals.—*Boston Cooking School.*

(25)

FISH

MEATS

ENTREES

SAUCES

SALADS

Desserts

CAKES

ICES

MISC.

FISH PUDDING.

1 pound boiled fish.	½ cup of cream.
1 ½ tablespoons of flour.	1 ½ teaspoons of salt.
¼ teaspoonful of pepper.	1 teaspoonful lemon juice.
A little onion juice.	2 eggs.

Mash the fish thoroughly, then put through a puree sieve and add seasonings. Put butter in the sauce pan, and when melted add the flour, then the cream, then the beaten eggs, stirring until well scalded, not thick. Then add the fish, beat well and fill a ring mold with the pudding, pressing it well against the sides ; set the whole in a pan of water and put in a moderate oven for thirty minutes. Remove onto a dish and fill in the center with parisienne potatoes, making a border of the same outside, and serve with rich cream sauce, in which parsley is chopped.— *Century Cook Book.*

FISH CROQUETTES.

1 pint boiled fish.	½ teaspoonful of onion
½ teaspoonful pepper.	juice.
1 cupful of cream.	1 tablespoonful of butter.
2 tablespoonsful of flour.	Yelks of two eggs and a
1 teaspoonful salt.	little chopped parsley.

Put the butter into a sauce pan ; when melted add the flour, and when thoroughly mixed add the cream, then the seasonings, then the beaten yelks of two eggs, and then the fish and the parsley. Spread on a dish to cool ; make out into croquettes ; to the beaten yelk of one egg add two tablespoonsful of water. Dip the croquettes first into the stale bread crumbs, then in egg, and then in crumbs. Fry in boiling fat. Serve with either Bechamel or Hollandaise sauce.

LOBSTER CUTLETS.

2 level tablespoonsful of butter.	1 pint chopped lobster.
1 teaspoonful of salt.	2 heaping tablespoonsful of flour.
½ cup of white stock.	Pepper to taste.
½ cup of cream.	1 beaten egg.
1 teaspoonful finely chopped parsley.	Mushrooms and 1 chopped truffle.

Melt butter, flour, cream, and stock. Work smooth ; add the parsley, egg, lobster, mushrooms, and truffle. Cook a few minutes and pour out on a platter to get thoroughly cold (the colder the better) ; shape into cutlets, dip in bread crumbs, then in diluted egg, then in crumbs ; put in frying basket and fry in hot lard.

FRIED OYSTERS.

Take large, select oysters, wash and drain and wipe. Dip them in the yellow of an egg, diluted with two tablespoonsful of water, then in bread or cracker crumbs; put in frying basket and fry in deep, hot lard.

OYSTERS EN COQUILLE.

2 sets of calf brains.	50 oysters.

Carefully clean the brains and boil in salt water ; scald the oysters in their own liquor until the edges curl, and then cut in small pieces. Chop the brains and mix with the oysters. Take two tablespoonsful of butter and saute a little finely chopped onion in it, add to the brains and oysters a little chopped parsley, celery salt, salt and pepper. Then add one-half cup of cream, two tablespoonsful of stale bread crumbs,

MEATS

ENTREES

SALADS SAUCES

Desserts

CAKES

ICES

MISC.

and oyster liquor to moisten.   Serve in shells or
bread sticks, with buttered bread crumbs on the top
of each; put in the oven just long enough to get
thoroughly hot.

STUFFED LOBSTER.

2 pounds lobster meat.	1 ½ cups of cream.
½ cup rich white stock.	A bit of bay leaf.
3 tablespoonsful of butter.	3 tablespoonsful of flour.
Yelks of 2 eggs.	1 teaspoonful lemon juice.
1 teaspoonful of chopped parsley.	1 cup of pecan kernels (if desired).

Season with salt, cayenne, and a little grated nutmeg. Scald stock and cream with the bay leaf, remove
the bay leaf, melt the butter ; then add the flour,
then cream and stock, and then the seasonings, then
the yelks slightly beaten, and the lemon juice. When
sauce is thick, add the lobster and the nuts, and fill
the shell ; cover with buttered bread crumbs and
brown in the oven.   Serve in a nest of watercress.

LOBSTER A LA NEWBURG.

1 pint finely chopped lobster.	½ pint cream.
Yelks of 3 eggs.	⅓ glass of sherry.
½ teaspoonful of salt.	A little red pepper.

Put the cream, wine, and beaten yelks together
in a double boiler and cook, stirring steadily, until
the sauce thickens.   Put in the lobster, let it become
heated through, season and serve.   A larger portion
of sherry may be used if desired.   Be very careful
to cook this over boiling water, as it curdles very
easily.

MEATS.

ROAST FILLET.

The fillet should be plentifully larded and all of the sinewy skin and gristle removed from the top, and most of the fat from the under side. Then place in a baking pan thin slices of larding or pickled pork, chopped onion, carrot, turnip, and celery ; then place the fillet on this. Pour over it a cupful of brown stock, salt and pepper, chopped parsley, bay leaf, and cloves. Cook in a hot oven for thirty minutes, basting frequently. When done, drain off the gravy and remove grease from the top. Take a tablespoonful of butter, add a tablespoonful of flour, cook together until they are brown. Add the gravy and a little brown stock—a cupful in all—stir until it boils, add a canful of mushrooms, chopped, and let it simmer for five minutes ; then add a little Madeira or sherry ; pour round the fillet and serve.

BROILED FILLETS.

Select small beef tenderloins, two inches thick ; lard thoroughly ; let them lay for two hours in a strong, highly seasoned stock with two tablespoonsful of claret ; broil for a few minutes over a hot fire ; serve with drawn butter or mushroom sauce.

3   (29)

BEEFSTEAK IN OYSTER BLANKET.

Select a porterhouse steak at least an inch and a half thick, remove the bone, wipe off with a wet cloth, rub over with lemon. With some of the fat which must be trimmed off, grease the wire broiler and place the meat in it; broil over a very hot fire at first that the surface may be well seared, thus preventing the escape of juices. After this, turn occasionally until cooked on both sides; remove to a baking pan, cover thoroughly with select oysters, placing a little butter here and there all over it. Squeeze the juice of half a lemon and place in a hot oven; cook until the oysters plump and the edges curl ; season with salt and pepper. Serve with melted butter, a little lemon juice, and chopped parsley.

STUFFED SHOULDER OF MUTTON.

Have the butcher carefully remove the blade from the shoulder, and fill the space with a mixture made of one cupful of crumbs, one tablespoonful of butter, one tablespoonful of chopped parsley, one dozen oysters, the juice of one lemon, a teaspoonful of salt, and one egg. Sew up the opening, roast in the oven with a little water in the pan ; allow fifteen minutes to the pound, and baste frequently. More oysters may be used, or they may be omitted altogether. A stuffing may be made of chopped meat, celery, onion, mushrooms, crumbs, egg, and seasoning of salt and pepper.—*Century Cook Book.*

STUFFED LAMB CHOPS.

Put a tablespoonful of butter into a saute pan ; when hot, add a tablespoonful of flour ; let the flour cook a few minutes, add four tablespoonsful of chopped mushrooms, one teaspoonful of parsley, one-half teaspoonful of salt, and a dash of pepper. Moisten with three tablespoonsful of stock ; mix all together and set aside to cool. Have six French chops cut one and a half inches thick, then split them in half, cutting to the bone ; spread the mixture between the split chops, press the edges well together, and broil eight minutes. Serve with melted butter or Spanish sauce.

GOOD STUFFING FOR TURKEY OR CHICKEN.

Moisten a cupful of bread crumbs with melted butter, season highly with salt, pepper, thyme, chopped parsley, and onion juice. Or, put in a sauce pan a tablespoonful of butter and fry in it one onion chopped fine, then add a cupful of bread which has been soaked in water, all of the water having been pressed out thoroughly, one-half cupful of stock, a teaspoonful of salt, a teaspoon each of pepper and thyme, one-half cup of celery cut into very small pieces. Stir it until it leaves the sides of the pan, then stuff either turkey or chicken.—*Century Cook Book*.

ENTREES.

CHICKEN CROQUETTES.

1 boiled chicken.   1 can mushrooms.
½ pound stale bread.   ¼ pound butter.
4 eggs.

Put the chicken and mushrooms through a croquette grinder. Soak the stale bread in a little chicken broth and add it to the meat and mushrooms, then add the butter and eggs; mix well together and boil until well cooked; season with salt, pepper, celery salt, chopped parsley, and a little finely chopped onion, and a very little nutmeg. Pour out on a platter, and when thoroughly cold, shape, roll in bread crumbs, place in a frying basket and fry in boiling fat.

SUPREME OF CHICKEN.

Breast and wing of a four pound chicken, raw. Four eggs, two-thirds cup of thick cream. Force the chicken through the meat grinder, beat the eggs separately and add, stirring until the mixture is smooth; add the cream, salt, pepper, and a little celery salt. Butter the timbal molds and line them with chopped mushrooms ; fill with the chicken and set the molds in a pan of boiling water. Bake about thirty minutes in a moderate oven. Serve with Bechamel sauce.

ENTREES

SALADS SAUCES

Desserts

CAKES

ICES

MISC.

BOUDINS OF CHICKEN.

For every pint of chopped, cooked chicken meat take one tablespoonful of butter, two of bread crumbs, one-half cup of cream, two whole eggs, one tablespoonful of chopped parsley, salt and pepper to taste. Melt the butter and add the bread crumbs; stir until well mixed, and add the cream. As soon as it is heated, take from the fire, add the chicken, seasoning, and the eggs beaten light without separating. Stir all well together, fill the cups or tins two-thirds full of mixture, stand in a baking pan half full of boiling water, and bake in a moderate oven about twenty minutes.

CHICKEN KLOPPS.

2 cups finely chopped cold chicken.	Unbeaten whites of four eggs.
1 teaspoonful of salt.	¼ teaspoon of Paprica.

Mix the above ingredients thoroughly ; moisten the hands with cold water and shape the mixture in round balls. Have ready a sauce pan of white stock just at the boiling point; carefully put the balls into it and poach about five minutes without allowing the water to bubble. Serve on rounds of buttered toast with a stalk of asparagus in each klopp. Pour around them a rich, white sauce.

SWEETBREAD A LA DIPLOMAT.

Saute half a tablespoonful of chopped onion in two tablespoonsful of butter, add two tablespoonsful of flour, half a cup of cream, and half a cup of white stock. Season with salt and cayenne. Add the yelk of one egg. When it thickens, add one third of a cup of mushrooms, chopped fine, two tablespoonsful chopped truffles, trimmings of the sweetbreads, and a little chopped parsley. Then add one tablespoonful of sherry wine. Let it cool and spread on sweetbreads which have been sauted in butter, after parboiling. Dip in diluted egg and bread crumbs, place in frying basket and fry in very hot fat. Serve with Allemande sauce.

SWEETBREAD A LA VICTORIA.

Boil two pairs of sweetbreads in salt water with a tablespoonful of lemon juice ; drain and cover with cold water. When cold, chop fine and add sufficient chopped mushrooms to make one pint in all. Melt one tablespoonful of butter, add one tablespoonful of flour, salt and pepper to taste. When smooth, add slowly a cup of cream. When this thickens, add a tablespoonful of lemon juice and a slight grating of nutmeg, half a teaspoon of finely chopped parsley, one beaten egg, and the sweetbreads and mushrooms. Pour out on a dish to cool, make palm shape, roll in bread crumbs, diluted egg and bread crumbs, place in frying basket and fry in hot lard. Serve with Allemande sauce.

SWEETBREAD CROQUETTES.

2 pairs sweetbreads.	A few chopped mushrooms.
1 level tablespoonful of butter.	1 heaping tablespoon of flour, salt, pepper, and
1 cup of cream.	a little onion juice.

Parboil the sweetbreads, putting a little lemon juice in the water. Throw them into cold water:

SAUCES

SALADS

Desserts

CAKES

ICES

MISC.

remove the outside skin and membrane. Chop fine and measure. Add enough chopped mushrooms to make a pint. Melt the butter and add the flour and then the cream. When smooth, add the yelk of one egg. Season with salt, pepper, and a little onion juice, chopped parsley, and celery salt. Then add the sweetbreads and mushrooms. Cook a few minutes, turn out to cool, shape, dip in bread crumbs, diluted eggs and crumbs, place in a frying basket and fry in hot lard.

STUFFED PEPPERS.

Cut off the tops of green peppers and remove the seed, parboil them ten minutes, chop the tops fine, one tablespoonful of chopped onion and two of fresh chopped mushrooms; saute all in two tablespoonsful of butter about twelve minutes, add one tablespoon of flour, half a cup of brown stock, one tablespoon of ground chicken, one-half tablespoon of ground ham, and one tablespoon of bread crumbs. Season with salt, pepper, and chopped parsley, cool, then stuff the peppers, sprinkle with buttered bread crumbs and put in the oven to brown; serve with white sauce.

PEPPER TIMBALS.

Butter well a tin timbal mold or cup, line with a large red pepper from which has been taken the seeds (and which has been parboiled, or use the canned red pepper), butter them and line with chopped mushrooms. Drop into each one a raw egg, sprinkle over a little salt and pepper, put into a baking pan which is half full of boiling water, and put into a hot oven and cook until the egg is thoroughly done. Turn out and serve with white sauce.

EGGS A LA TURK.

Brown one chicken liver and one large mushroom together in butter one minute. Add a little chopped onion, salt and pepper, and a tablespoonful of flour; beat until smooth. Then add one tablespoon of sherry and enough brown stock to make a sauce— about half a cupful—one teaspoonful of lemon juice, and a few chopped truffles. Place a poached egg, well cooked, on round buttered toast, and serve the sauce around it.

MUSHROOMS A L'ALGONQUIN.

Wash, peel, and remove the stems from large selected mushrooms, and then saute in butter ; when done, put in a buttered pan, placing on each a large oyster ; sprinkle with salt and pepper, place on each a bit of butter, cook in a hot oven until the oysters are plump. Serve with drawn butter sauce.

SAUCES.

ALLEMANDE SAUCE.

Melt two level tablespoonsful of butter, and add two heaping tablespoonsful of flour ; when smooth, pour on one-half pint of white stock and one-half pint of cream; season with salt, pepper, chopped parsley, and lemon juice, and then add the beaten yelk of an egg.

BECHAMEL SAUCE.

1 ½ cups white stock.	1 slice of onion.
1 slice of carrot.	1 bay leaf.
1 sprig of parsley.	¼ cup of butter.
¼ cup of flour.	1 cup of cream.
Salt and pepper.	

Cook the stock with the onion, carrot, bay leaf, and parsley about fifteen minutes, and then strain. Melt the butter, add the flour, then the stock and cream.

LOBSTER SAUCE.

2 tablespoonsful of butter.	· 2 tablespoonsful of flour.
1 pint of cream.	Yelks of two eggs.

Season with salt, pepper, and a little sherry wine. Melt the butter, then add the flour, then the cream, then the seasoning, and then the well beaten yelks, and when thick add a heaping cup of lobster, chopped fine.

(39)

HOLLANDAISE SAUCE.

½ cup of butter.   ½ teaspoon of salt.
Yelks of 4 uncooked eggs.   ⅓ cup of boiling water.
1 ½ tablespoonsful lemon   Dash of cayenne.
  juice.

Fill a bowl with hot water, pour out the water and wipe the bowl dry. Put the butter into it and beat until soft and creamy ; add the yelks of the eggs, one by one, and beat until they are blended with the butter. Add the lemon juice, salt, and pepper, and beat again until smooth. Then take out the spoon and beat the mixture with an egg-beater five minutes. Put into a double boiler with boiling water. Add to the butter and eggs one-third cup of boiling water and cook until the same is as thick as mayonnaise, beating constantly with the egg-beater. Serve either hot or cold.

MUSHROOM SAUCE.

1 tablespoonful of butter.   1 heaping tablespoon of flour.
¾ cup brown stock.   ¼ cup of cream.

Season with salt and pepper, and add one cup of chopped mushrooms.

HORSERADISH SAUCE.

Mix two tablespoonsful of grated horseradish with one tablespoonful of vinegar and one-fourth teaspoonful each of salt and pepper. Mix thoroughly and stir in four tablespoonsful of whipped cream, stiff. Serve with roast beef or oysters.

MAYONNAISE No. 1.

Yelk of 1 hard boiled egg.	1 teaspoonful of mustard.
Salt and pepper to taste.	Yelk of 1 raw egg, well
½ of small bottle olive oil.	beaten.
White of 1 egg beaten stiff	Vinegar to taste.
and dry.	

Rub the yelk of hard boiled egg through a fine sieve until smooth ; add to that the mustard, salt, pepper, raw yelk, well beaten. Then add the oil, and next the vinegar slowly, lastly the raw white of egg.

MAYONNAISE No. 2 (COOKED).

Yelks of two eggs, well beaten, four tablespoonsful of vinegar. Boil until thick, and stir in one heaping tablespoonful of butter or olive oil. When cold, add half a teaspoonful of salt, half a teaspoonful of dry mustard, and a little pepper, and a cup of whipped cream.

GARIBALDI SAUCE.

1 pound tart apples, pared and cored.	4 ounces onions.
	1 pound salt.
1 pound ripe tomatoes, cut fine.	½ pound best ginger, ground.
1 pound best layer raisins.	2 ounces garlic.
½ pound peppers bell, ripe but not dried.	

Chop all ingredients fine, and put in a stone jar, adding four quarts of best vinegar ; let stand four weeks, stirring daily. Then boil thirty minutes, strain through a colander and then through a sieve, and bottle when cold.

SALAD CREAM.

Mix one-half tablespoonful of mustard and salt (each) and one tablespoonful of sugar with one egg, slightly beaten. Pour on this three-fourths cup of cream and one-fourth cup of scalded vinegar with two and a half tablespoonsful melted butter. Cook in a double boiler until it thickens slightly. Strain and cool. Serve on cold slaw.

GERMAN DRESSING.

Beat one-half cup of heavy cream, just beginning to sour, with one egg ; beat until stiff. Add three tablespoonsful of vinegar and beat again.

SALADS.

CHICKEN SALAD.

Take equal proportions of cold chicken and celery, cut not too small. To a quart of chicken and celery pour over one-half cup of oil and let it marinate half or whole morning, and when ready to serve mix with mayonnaise dressing.

SALAD A LA JARDINE.

To one pint of ground, boiled chicken, add equal parts of asparagus tips, peas, chopped string beans, chopped celery, and a few pecan kernels. Mix carefully and pour over it mayonnaise.

EGG SALAD.

Boil the eggs twenty minutes. Peel off the shells and cut the eggs in half lengthwise. Remove the yelks, put in a bowl and cream. Take two eggs well beaten, half a teaspoonful of dry mustard, three tablespoonsful of rich, sweet cream, one tablespoonful of salt, one teaspoonful of pepper, two tablespoonsful of olive oil, and one and a half tablespoonsful of vinegar. Boil all until very thick and mix with the cooked yelks. Fill the whites, and when cold serve with mayonnaise.

(43)

NUT AND CELERY SALAD.

Mix equal parts of pecans, almonds, English walnuts, and celery. Marinate in oil and serve with a French dressing with a border of curly celery and lettuce.

FROZEN TOMATO SALAD.

Take one quart can of tomatoes (or the same proportion of fresh tomatoes), drain off all the liquor, pour over them mayonnaise and a little chopped celery, put in a freezer and freeze. Serve in nasturtium leaves.

TOMATO JELLY.

Cook one-half can of tomatoes for ten minutes, with a pinch of soda if very acid. Add half a teaspoonful of salt, and rub through a sieve or strainer. Pour over it one-fourth box of gelatine which has been soaked in one-fourth cup of cold water; mold, and when congealed serve on lettuce with mayonnaise dressing.

ORANGE SALAD.

Select firm, acid oranges; cut in half and remove all the pulp from the skins. Marinate in two tablespoonsful of oil, one tablespoonful of lemon juice, and a little salt. Make a dressing with half a cup of whipped sour cream, one tablespoonful of lemon juice, one-fourth teaspoonful of salt, and a little cayenne. Remove oranges from the oil, place back in the skins, pour over the cream dressing, and serve on lettuce leaves.

GREEN GRAPE SALAD.

Select firm, acid grapes ; serve in a head of lettuce with the cooked mayonnaise, only with a little more cream added to it, or with a cream dressing, for which mix half a teaspoonful of salt, half a teaspoonful of mustard, one-fourth teaspoonful of sugar, one egg beaten slightly, two tablespoonsful of oil, three-fourths cup of rich cream, and a scant quarter of a cup of vinegar.

4

DESSERTS.

ORANGE PUDDING.

1 pint of cream.
½ cup of sugar.
2 tablespoonsful corn starch.

Whites of 4 eggs—pinch of salt.

Dissolve the corn starch in a little of the cream; put the remaining cream and sugar over the fire in a double boiler. When it boils, add the corn starch, cook until smooth, then add eggs well beaten. When cold, slice six oranges fine, sprinkle with sugar, and let them stand for half an hour. Pour over foundation, and over this pour whipped cream and oranges, just before serving.

ENGLISH PLUM PUDDING.

½ pound stale bread crumbs.
¼ pound sugar.
½ nutmeg, grated.
½ teaspoonful mace and ground cloves.
1 teaspoonful salt.
1 ½ pounds raisins.
¼ pound figs.

½ cup wine and brandy mixed.
1 cup hot milk.
1 teaspoonful cinnamon.
4 eggs.
½ pound beef suet.
¼ pound currants.
⅛ pound citron.

Soak the stale bread crumbs in one cup of hot milk. When cold, add the sugar and yelks of eggs beaten stiff, also nutmeg, cinnamon, mace, ground cloves, and salt. Chop fine and cream the beef suet

(47)

and add to the mixture with the raisins stoned and floured, and the currants, figs, and citron chopped fine.   Add the wine and brandy, and the whites of four eggs beaten stiff.   Turn into a buttered mold and steam from six to eight hours. — *Boston Cooking School.*

RICE PUDDING.

4 tablespoonsful of rice.	Milk and cream.
½ teaspoonful of salt.	4 tablespoonsful sugar.
1 teaspoonful vanilla.	½ cup of stoned raisins.

Into a pudding dish holding a quart put the rice, which has been well washed and soaked.   Fill the dish with milk and cream, and add the salt.   Put into the oven to cook for about half an hour.   Add the sugar, vanilla, and raisins, and return to the oven and cook slowly for two hours or more if necessary. If the milk boils down, lift the skin at the side and add a little more hot cream.   To make the pudding creamy it must be cooked very slowly and plenty of cream used.   Just before serving, spread thickly over the top fresh marshmallows.   Put in the oven just long enough for the marshmallows to swell.   Before sending to the table, garnish with candied cherries or red jelly.   Served with whipped or plain cream.

MINCEMEAT PATTIES.

Heat pattie shells and mincemeat separately. When very hot, fill the shells with the mincemeat and serve with frozen whipped cream, flavored with brandy.

FIG PUDDING.

⅓ pound beef suet.
2 heaping cups stale bread
crumbs.
½ cup milk.

1 teaspoonful salt.
½ pound figs.
2 eggs well beaten.
1 cup sugar.

Chop and rub to a cream the beef suet, add the raisins finely chopped ; mix thoroughly. To the bread crumbs add the well beaten eggs, milk, sugar, and salt, and mix all together well. Place in a buttered pudding dish and steam for several hours. Serve with a fancy sauce.

SAUCE FOR SAME.

Beat the yelks of two eggs until light. Then beat the whites of two eggs stiff and add half a cup of powdered sugar. Combine the two and add one-fourth cup of hot cream and four tablespoonsful sherry wine.—*Boston Cooking School.*

CHARLOTTE RUSSE.

White of one egg.
⅓ box gelatine dissolved in
½ pint boiling water.
Yelks of three eggs.

Sponge lady fingers.
1 cup powdered sugar.
2 teaspoonsful vanilla.
Whip from 1 qt. of cream.

Beat the white of an egg slightly, put a thin coating around a glass bowl, and then line with sponge lady fingers. Dissolve the gelatine in boiling water. When thoroughly dissolved, stir in the sugar, add the vanilla and the beaten yelks of three eggs; stir in the whip from a quart of cream, and when it stiffens some, pour into the bowl lined with sponge cakes and garnish the top prettily with whipped cream.

CABINET PUDDING.

1 pint of milk.	½ cup raisins, chopped
2 tablespoonsful of sugar.	citron, currants.
½ tablespoonful of butter.	¼ teaspoonful of salt.
2 eggs.	1 ½ pints stale sponge cake.

Beat the eggs, sugar, and salt together ; add the milk ; sprinkle a pudding mold with cake crumbs, then a layer of fruit, then cake crumbs, and continue until all is used up.   Pour on the custard and let it stand two hours, then steam one and a half hours.

SAUCE FOR SAME.

1 cup of butter.	2 cups powdered sugar.
½ cup of cream.	

Beat the butter to a cream, add the sugar gradually, and when very light, add the cream.   Flavor to taste.   Cook for a few minutes in a double boiler.

PLAIN PASTRY.

1 ½ cups flour.	¼ cup of butter.
¼ cup lard.	½ teaspoonful salt.
A little ice water.	

Wash the butter, squeeze out all the milk and water, flatten it out.   Add the salt to the flour and cut in the lard with a knife.   Moisten it with the cold water.   Toss on the board, dredged sparingly with flour, pat and roll out.   Fold in the butter, roll out, and repeat folding and rolling several times. Cover with cheese cloth and set away in a cool place, though never in direct contact with ice.   Roll thin and bake in a moderate oven.—*Boston Cooking School.*

SIMPLE DESSERT.

Lady fingers.
½ cup of sherry wine.
Whipped cream.

Bananas, sliced thin.
2 heaping tablespoonsful
of sugar.

Line a bowl with lady fingers, fill it half full of bananas sliced thin, pour over them about half a cup of sherry wine and a heaping tablespoonful of sugar, then fill the bowl with whipped cream.

WINE JELLY.

4 ounces of sheet gelatine.
2 pounds of sugar.
1 quart of sherry wine.
5 pints of water.

1 medium tumbler of lemon
juice.
Whites of 3 eggs.

Pour the water on the gelatine, soak over night, and in the morning put all of the ingredients together, including the beaten whites and shells of the eggs. Boil twenty minutes ; add the thin yellow rind of two lemons five minutes before taking off the fire. Strain through a canton flannel bag, pouring back the first that runs out. Do not squeeze, but let it drip until all runs out. Pour in molds and set away to congeal.

BAVARIAN CREAM.

⅓ box of gelatine.
Sweeten and flavor to taste.

½ cup of boiling water.
1 quart of whipped cream.

Soak the gelatine in the boiling water, sweeten and flavor to taste ; add one quart of stiff whipped cream ; put in molds and set away to congeal, and serve with whipped cream.

BAKED CARAMEL CUSTARD WITH SAUCE.

Set a small sauce pan, containing one-half cup of sugar, over the fire and stir the sugar gently. As the sugar loses water by evaporation it assumes the appearance of flake tapioca, and as the cooking continues it changes color, becoming caramel. Care must be taken that the caramel does not burn or become too dark in color. Scald four cups of milk, and add the caramel to the milk very carefully, and as soon as the two are well blended, pour the mixture onto five eggs slightly beaten ; then add one-half teaspoonful of salt and one teaspoonful of vanilla. Strain at once into a buttered melon mold, set the mold in a pan of hot water and bake in a slow oven until the custard is firm. Serve with caramel sauce.

SAUCE FOR SAME.

Put one-half cup of sugar into a sauce pan over the fire and stir the sugar until it melts and becomes a light brown color. Add half a cup of boiling water, and allow the liquid to simmer ten minutes.—*Miss Farmer.*

PINEAPPLE PUDDING.

2 ¾ cups of scalded cream.   ¼ cup of sugar.
⅓ cup of corn starch.   ½ can grated pineapple.
¼ teaspoonful of salt.   Whites of three eggs,
¼ cup of cold milk.   beaten stiff.

Mix the corn starch, sugar, salt, and cold milk well, and add to the scalded cream in a double boiler, stirring constantly until it thickens. Cook from ten to fifteen minutes, add the eggs, then pineapple. Mold, congeal, and serve with whipped cream.

FILLING FOR LEMON PIES.

Juice of 2 lemons.	2 cups of sugar.
2 cups boiling water,	3 tablespoonsful flour.
1 tablespoonful butter.	Yelks of 4 eggs, well beaten.

Boil all together until very thick. Fill the cooked crusts, and use the whites for meringue on top.

COCOANUT FILLING FOR PIES.

½ cup butter.	2 cups of sugar.
Yelks of 5 eggs.	1 cup milk.
1 tablespoonful flour.	1 cup grated cocoanut.

Flavor with vanilla and cook until thick. When cold, fill the cooked pie crusts and cover with meringue; put into the oven to brown.

PRESERVES IN HALF ORANGES.

Take half of an orange, scoop out all of the pulp, cut the edge in points, fill in with preserves—pineapple being prettiest—and serve with whipped cream.

CAKES.

WHITE LADY CAKE.

12 eggs.
1 teacup of butter.
½ cup of cream.
2½ teacups sugar.

3½ teacups of flour.
3 teaspoonsful baking powder.

Cream the butter and sugar together until very light ; add the whites of eggs beaten stiff, then the flour, and then the baking powder stirred in the cream. Bake in a solid cake in a moderate oven for very nearly one hour. Any desired flavoring may be used.

LAYER CAKE.

1 cup butter.
3 cups flour.
½ cup milk.
2 cups sugar.

6 eggs.
2 heaping teaspoons baking powder.

Take only the whites of eggs, beaten stiff. Mix as in Lady Cake, and bake in tins in a moderate oven.

SPONGE CAKE.

12 eggs.
1⅓ cups flour.
1¼ cups of sugar.

1 level teaspoonful cream tartar.

Beat the yelks of eight eggs with the sugar until very light. Beat the whites of twelve eggs with the cream tartar to a stiff froth. Add to the yelks and sugar, then add the flour slowly ; flavor to taste, and bake in a moderate oven forty minutes.

(55)

ANGEL FOOD.

1 2 eggs, whites beaten stiff.	1 teaspoonsful cream tar·
1 ½ tumblers powdered	tar.
sugar.	1 tumbler of flour.

Take the whites of eggs beaten to a stiff froth with the cream of tartar added. Sift the powdered sugar into the eggs and cut it in with an egg-beater (never stir Angel Food with a spoon). After the flour has been sifted five times, sift very slowly into the egg and sugar. Add a teaspoonful of vanilla. Grease the cake pan very little with butter, lining the bottom with unglazed letter paper which has been slightly greased. Pour in the cake and bake forty minutes. Put a pan of water over it from the first. Remove from the oven, invert the pan, and let it stand until the cake falls out without being disturbed.

HICKORY NUT CAKE.

½ cup butter.	1 cup granulated sugar.
3 eggs.	1 cup milk.
1 ½ cups flour.	1 ½ teaspoons baking pow·
1 cup hickory nuts chopped	der.
fine.	

Cream the butter and add the sugar gradually. Beat the yelks of three eggs light and add to the butter and sugar with one cup milk. To the flour add the baking powder, stir into the batter, add the hickory nut meats chopped fine, and the whites of two eggs beaten stiff. Bake in a buttered and floured pan from forty to fifty minutes, or in small pans.

PECAN CAKE.

1 pound sugar.	1 pound flour.
1 pound of butter.	10 eggs.
½ tumbler of brandy.	2 grated nutmegs.
1 pound of raisins.	¼ pound of citron.
1 ½ pounds of pecan kernels.	

Cream the butter and sugar until light. Add the eggs beaten separately, then the nutmeg stirred in the brandy, then the flour, raisins, citron, and pecan kernels. Pour into buttered mold and bake half an hour longer than you would a black cake, same size.

SPICE CAKE.

1 cup of sugar.	½ cup of sour milk.
½ cup of butter.	1 teaspoonful of soda.
2 ½ cups of flour.	1 tablespoonful ginger.
1 tablespoonful cinnamon.	½ teaspoonful of cloves.
4 eggs.	1 ½ pounds raisins (if de-
½ teaspoonful allspice.	sired).
½ cup of molasses.	

Use only the well beaten yelks of eggs. Bake in small pans, or as a solid cake.

GINGER BREAD.

1 cup molasses.	⅓ cup of butter.
1 ¾ teaspoonsful soda.	½ cup sour milk.
1 egg.	2 cups flour.
2 teaspoonsful ginger.	½ teaspoonful of salt.

Put the butter and molasses in a sauce pan and cook until the boiling point is reached. Remove from the fire, add the soda and beat vigorously, then add the milk, eggs well beaten, and the remaining ingredients mixed and sifted. Bake fifteen minutes in buttered pans two-thirds filled with the mixture.
—*Miss Farmer.*

SOUR MILK GINGER BREAD.

1 cup molasses.	1 cup sour milk.
2⅓ cups flour.	1¾ teaspoonsful soda.
2 teaspoonsful ginger.	½ teaspoonful salt.
¼ cup melted butter.	

Add the milk to the molasses, mix and sift the dry ingredients, combine the two, add butter and beat vigorously. Pour into a buttered, shallow pan and bake twenty-five minutes in a moderate oven.

CRULLERS.

2 cups of butter.	3½ cups of sugar.
12 eggs.	Flour enough to roll.

Flavor with nutmeg or cinnamon, roll thin, shape and fry in hot fat.

SAND TARTS.

1 cup butter.	2 cups sugar.
3 eggs.	Flour enough to roll.

Roll thin, paint the tops with the white of egg, sprinkle over with equal parts of ground cinnamon and granulated sugar, and in the center of each place one fourth of a blanched almond. Put in floured pans and bake in a quick oven.

COOKIES.

3 cups sugar.	1½ cups butter.
6 eggs.	5 pints flour.
3 teaspoonsful carbonate of ammonia.	

Cream the butter and sugar, beat the eggs three at a time into it, and then beat well. Add the ammonia, and, lastly, flour and roll thin.

FILLINGS FOR CAKES.

PLAIN CARAMEL.

2 cups of sugar.	¾ cup of maple syrup.
Cream to wet thoroughly.	1 tablespoonful butter.

Put sugar, syrup, and cream on, and when it boils add the butter. Boil it until very thick. Add one teaspoonful of vanilla, take from the fire, and beat until it begins to sugar. Then pour over the cake.

CHOCOLATE CARAMEL.

Same as above, only before it begins to boil add one-fourth cake of Baker's chocolate.

ICE CREAM FILLING.

3 cups sugar.	1 cup water.
3 eggs, whites beaten stiff.	1 teaspoonful vanilla.

Boil sugar and water to a candy, pour slowly over the beaten whites of three eggs, flavor with vanilla, beat until it begins to cream, and pour over the cake.

MARSHMALLOW AND PINEAPPLE FILLING.

Take fresh marshmallows, put into the oven to soften, spread over the cake with a little chopped, candied pineapple, and pour over same the Ice Cream Filling given above.

ICES.

MISC.

CREAM ICING FOR ANGEL FOOD.

3 cups of sugar.   1 cup of cream.
½ teaspoonful of vanilla.

Let it come to a good hard boil, beat hard until creamy, and pour over the angel food.

PRAULINE ICING.

Make a plain caramel, and when done, add one cup of broken pecan kernels just before pouring on the cake.

ICES.

NESSELBRODE PUDDING.

1 cup of marons.	1 cup of granulated sugar.
Yelks of 3 eggs.	½ pint of cream.
¼ pound of candied fruits.	½ can pineapple (drained).

Take candied fruits and marons and soak them in sherry wine. Put sugar on the fire with one fourth of a cup of boiling water and boil to a syrup. Beat the yelks of eggs until light. Pour on them slowly the syrup, stirring all the time. Put on the fire in a double boiler and cook until the consistency of thick cream. Remove and beat hard until cold. When cold, add the cream, the marons pounded, and half a teaspoonful of vanilla, and freeze. When nearly hard frozen, add the candied fruits, one fourth of a pound of raisins, one fourth of a pound of pounded almonds, and a glass of sherry wine, and freeze hard. Remove the dasher and allow it to stand for several hours.—*Century Cook Book.*

PLAIN VANILLA CREAM.

Take one quart of plain, rich cream, season and flavor. When half frozen, add one quart of stiff whipped cream which has been sweetened and flavored. Freeze hard. Pack for an hour before using.

5   (61)

TO FREEZE A WATERMELON.

Take three pints of stiff whipped cream, color with Burnett's Green Vegetable Coloring, sweeten and flavor with extract of pistachio, put in a freezer and freeze very hard.

Then take a quart of very stiff whipped cream, sweeten and flavor with a little sherry wine, put in a freezer and freeze hard.

Then take a quart of stiff whipped cream, sweeten and color pink with Burnett's Vegetable Coloring, and flavor with strawberry. Put in a freezer and freeze hard.

Take a melon mold and line it with the green, then put a layer of the white, and then the pink, sprinkled well with Sultana Raisins that have been soaked in brandy, making the seeds. Cover with the white cream, and then the green ; put a piece of buttered letter paper over it and then the tin top. Pack in salt and ice, and let stand for several hours.

THREE OF A KIND.

Juice of 3 lemons.	Juice of 3 oranges.
Sugar to taste.	3 slices of canned peaches
2 bananas.	or pineapple.
1 quart of cold water.	

Take the lemon juice, cold water, and sugar, and a pint of rich cream—to be added after the lemon and water are packed in the freezer. When this begins to freeze, add the juice of three oranges, two bananas which have been put through a fine sieve, and three slices of canned peaches or pineapple put through a sieve. Freeze until very hard. Pack and serve.

HOLLANDAISE PUNCH.

4 cups of water.	1 ½ cups of sugar.
⅓ cup of lemon juice.	1 can pineapple.
¼ cup of brandy.	2 tablespoonsful of gin.

Cook the water, sugar, and a little grated lemon rind fifteen minutes. Add lemon juice and pineapple, cool, strain, and freeze partly, then add the liquor and continue freezing.

ORANGE ICE.

To four cups of sugar add a quart of water, and boil to a thick syrup. Add to this the juice of twelve oranges and four lemons, and one quart of cold water. Put in a freezer and freeze. Pineapple or any water ice may be made in the same way.

FRUIT PUNCH.

Take the same syrup as above ; add one quart of sherry, one-half pint of brandy, one-half pint of rum, one pound of candied cherries, one-half pound candied pineapple, half a pound of grapes, and the juice of six lemons with the extra quart of cold water.

MISCELLANEOUS.

POTATOES EN SURPRISE.

Season one pint of hot mashed potatoes with one tablespoonful of butter, one teaspoonful of salt, one-fourth teaspoonful of celery salt, one-fourth teaspoonful of pepper, and a few grains of cayenne. Add six drops of onion juice, cool slightly, and add the yelk of one egg beaten slightly. Shape into balls. Make a hole in the center, fill with creamed chicken, oysters, or sweatbreads. Close up, dip in crumbs, diluted egg and crumbs, and place in a frying basket and fry in hot fat. Serve with cream or oyster sauce.

PARISIENNE POTATOES.

With a French vegetable cutter cut potato balls out of peeled raw potatoes ; drop in cold water for about half an hour ; put in salted boiling water and boil about fifteen minutes, or until tender. Drain off the water, and let them stand on the back of the range, covered over, until dry. Serve with white sauce and chopped parsley.

LITTLE PIGS IN BLANKETS.

Select large, plump oysters, or firm pieces of sweetbread which have been parboiled. Wrap them in thin slices of fat bacon, pinning with a wooden toothpick. Broil in a little butter.

PASTRY CRULLERS.

1 quart flour.   2 cups water.
2 eggs.   1 tablespoonful of butter.

Mix the flour and water, then the butter, then
the beaten eggs and a little salt.   Have the cruller
iron heated thoroughly in boiling lard.   Be very
careful to drain all the lard from the iron, dip into
some of the batter which you have put into a pint
cup, being careful not to let the iron touch the bottom or sides of the cup ; then dip in boiling lard and
fry to a nice brown ; remove from the iron and heat
it again.   Serve plain this way as a garnish, or
sprinkle with cinnamon sugar as a cruller.

STUFFED TOMATOES.

Take firm, large tomatoes, not too ripe, cut out
the blossom end and scoop out the inside as clean as
you can without breaking the skins.   Chop fine and
add equal parts of ground chicken, chopped celery,
okra, and a few bread crumbs.   Season well with
salt and pepper and a little onion juice.   Fill the
skins, put a piece of butter on top of each, and place
in a buttered baking dish and bake in a good oven.

BAKED BANANAS.

Peel firm bananas and cut lengthwise ; place in a
baking dish.   Slice a lemon very thin, put a layer of
banana with three slices of lemon, and then a layer
of banana and three slices of lemon, sprinkled well
with sugar.   Put in the oven to bake.

FRICASSEED OYSTERS WITH MUSHROOMS.

30 oysters.	1 tablespoonful of flour.
1 tablespoonful of butter.	⅓ cup mushroom liquor.
1 cup cream.	1 teaspoonful of salt—
Yelks of 2 eggs.	little pepper.
½ cup mushrooms, sliced.	

Cook butter and flour together in double boiler. Pour on the cream and mushroom liquor, then the seasoning, and stir in the beaten yelks of egg very slowly. Add the oysters and mushrooms after the sauce is thick. Serve in pattie shells or on toast.— *Christine Terhune Herrick.*

CHICKEN WITH ASPARAGUS TIPS.

2 cups very tender chicken breast.	1 tablespoonful of butter.
1 cup cooked asparagus tips (fresh or canned).	½ pint cream.
	Yelks of 2 hard boiled eggs.

Rub the yelks and butter to a paste and add the cream. Stir until thoroughly blended. Season with salt and pepper ; then lay in the asparagus tips and chicken, and cook for a few minutes.

SALTED ALMONDS.

Blanch the almonds, wipe dry, place in a frying basket, then into Delmonico Cooking Oil, heated to the boiling point. When nicely browned, remove from the oil, sprinkle salt on them, and let them drain. Any other nut can be cooked in the same way.

CROQUETTES OF FRENCH PEAS.

2 tablespoonsful of butter.	2 tablespoonsful of flour.
1 pint of cream.	Yelk of one egg.
2 cans of peas.	Salt, pepper, and celery
1 teaspoonful onion juice.	salt.

Melt butter and flour together, then add the cream and seasonings, and the well beaten yelk of egg, and then the peas, which have been put through a puree strainer. Pour out onto a platter to cool, roll into croquettes, and fry as chicken croquettes.

EGG NOGG.

12 eggs.	12 tablespoonsful of sugar.
12 tablespoonsful best whisky.	12 tablespoonsful Jamaica rum.

Beat the yelks and sugar together until very light; then add the liquor slowly, next the whites, beaten to a stiff froth, and then one pint and a half of cream, whipped.

STUFFED SWEET POTATOES.

Select good, firm sweet potatoes, wash well and boil until tender.   Remove from the fire, cut in half lengthwise, take out most of the potato, leaving the skin firm enough to stuff. Mash the potato well, season with butter, cream, and a little sugar, cinnamon, and sherry wine to taste. Fill the skins with the mixture, and put in the oven to brown a little.

OMELET.

Experience has taught us that an omelet is the most difficult to prepare of all egg dishes. In the first place an omelet pan should never be used for any thing else. Before using, it is well to rub it with dry salt, to be sure it is perfectly smooth; and it is better to make several small omelets than to try to make one large one. Break from three to five eggs into a bowl, and beat twelve beats; sprinkle with salt and pepper and a few pieces of butter. Have the omelet pan hot, and put in just enough butter to cover the surface without being too greasy. Pour in the egg, and when it begins to cook carefully cut it in several places so that the uncooked egg may cook evenly. Then take a broad knife and fold it over, placing the dish on which the omelet is to be served on top the omelet pan; lift the pan carefully and turn out onto the dish. Garnish with parsley.

A FEW SIMPLE DISHES FOR THE SICK.

TOAST WATER.

Toast three slices of stale bread to a dark brown, but do not burn. Put into a pitcher, pour over them one quart boiling water. Cover closely and let stand on ice until cold. Strain. If desired, wine and sugar may be added.

RICE WATER.

Pick over and wash two tablespoonsful of rice. Put into a granite sauce pan with one quart boiling water. Simmer two hours, when rice should be softened and partially dissolved. Strain ; add a salt-spoonful of salt. Serve warm or cold. Two table-spoonsful of sherry or port may be added if desired.

BARLEY WATER.

Wash two ounces (one wineglassful) of pearl barley with cold water. Boil five minutes in fresh water. Throw both waters away ; pour on two quarts of boiling water and boil down to one quart. Flavor with thinly cut lemon rind. Add sugar to taste. Do not strain unless at the patient's request.

EGG WATER.

Stir the whites of two eggs into half a pint of ice water without beating the eggs. Add enough salt or sugar to make palatable.

FLAXSEED TEA.

Flaxseed, whole, 1 ounce   White sugar, 1 ounce.
(1 heaping tablespoonful).   Lemon juice, 4 tablespoons-
Licorice root, ½ ounce   ful.
(two small sticks).

Pour on these materials two pints of boiling water. Let stand in a hot place four hours and strain off the liquor.

PEPTONIZED MILK (COLD PROCESS).

In a clean quart bottle put one peptonizing powder (extract of pancreas 5 grains, bicarbonate of soda 15 grains—or the contents of one peptonizing tube—Fairchild), add one teacup of cold water and shake well. Add one pint of fresh cold milk and shake mixture again. Place on ice. Use when required without subjecting to heat.

PEPTONIZED MILK (WARM PROCESS).

Mix peptonizing powder with water and milk as described above; place bottle in water only so hot that the whole hand can be held in it a minute without discomfort. Keep the bottle there ten minutes. Then put on ice to check further digestion. Do not heat long enough to render the milk bitter.

PEPTONIZED MILK TOAST.

Over two slices of toast pour one gill of peptonized milk (cold process), let stand on the back of stove thirty minutes, serve warm or strain and serve fluid portion alone. Plain, light sponge cake may be similarly digested.

KOUMISS.

Take ordinary beer bottle with shifting cork, put in it one pint of milk, one sixth of a cake of Fleischmann's yeast, or one tablespoonful of fresh lager beer yeast (brewer's), one half of a tablespoonful of white sugar reduced to a syrup. Shake well and allow it to stand in the refrigerator two or three days, when it may be used. It will keep there indefinitely if laid on its side. Much waste can be saved by preparing the bottles with ordinary corks wired in position, and drawing off the koumiss with a champagne tap.

STERILIZED MILK.

Put the required amount of milk in clean bottles ; if for infants, each bottle holding enough for one feeding. Plug the mouths lightly with rubber stoppers, immerse to the shoulders in a kettle of cold water. Boil twenty minutes, or better, steam thirty minutes in ordinary steamer. Push in the stoppers firmly, cool the bottles rapidly, and keep in a refrigerator. Warm each bottle just before using.

BEEF TEA WITH ACID.

One and one-half pounds of beef from the round, cut in small pieces ; same quantity ice broken small. Let it stand in a deep vessel twelve hours ; strain thoroughly and forcibly through a coarse towel. Boil quickly ten minutes in a porcelain vessel. Let cool. Add one-half teaspoonful of acid, or acid phosphate, to a pint. Serve hot or cold.

MUTTON BROTH.

Lean loin of mutton, one and one-half pounds, including bone. Three pints of water. Boil gently until tender, throwing in a little salt and onion, according to taste. Pour out broth into basin ; when cold, skim off the fat. Warm up when wanted.

CHICKEN BROTH.

Chop up a small chicken, or half of a large fowl. Boil it, bones and all, with a blade of mace, a sprig of parsley, a tablespoonful of rice, and a crust of bread in one quart of water, for an hour, skimming it from time to time. Strain through a colander.

CREAM SOUP.

Take one quart of good stock, chicken or mutton ; cut one onion into quarters, slice three potatoes very thin and put into the stock with a small piece of mace. Boil gently for an hour. Then strain out the onion and mace. The potatoes should by this time have dissolved in the stock. Add one pint of milk, a very little corn flower to make it about as thick as cream, and a little butter. This soup may be made with milk instead of stock, if a little cream is used with it.

EGG LEMONADE.

Beat one egg with one tablespoonful of sugar until very light ; stir in three tablespoonsful cold water and the juice of a small lemon. Fill the glass with pounded ice and drink through a straw.

WINE WHEY.

Put two pints of new milk in a sauce pan and stir over a clear fire until nearly boiling. Then add one gill (two wineglassfuls) of sherry and simmer a quarter of an hour, skimming off the curd as it rises. Add one tablespoonful more of sherry and skim again for a few minutes. Strain through coarse muslin. May use two tablespoonsful of lemon juice instead of wine if desired.

JUNKET.

Take one-half pint of fresh milk, heated luke-warm. Add one teaspoonful essence of pepsin and stir just enough to mix. Pour into custard cups and let it stand until firmly curded. Serve plain or with sugar and grated nutmeg. May add sherry.

MILK AND EGG.

Beat milk with salt to taste. Beat white of egg until stiff. Add egg to milk and stir.

RUM PUNCH.

White sugar two teaspoonsful, one egg beaten up. Add a large wineglassful warm milk, two to four teaspoonsful Jamaica rum, and a little nutmeg.

CHAMPAGNE WHEY.

Boil one-half pint milk. Strain through cheese cloth and add one wineglass of champagne.

PEPTONIZED OYSTERS.

Mince six large or twelve small oysters. Add to them, in their own liquor, five grains extract of pancreas with fifteen grains bicarbonate of soda, or one Fairchild peptonizing tube. The mixture is then brought to a blood heat and maintained, with occasional stirring, at that temperature thirty minutes, when one pint of milk is added and the temperature kept up ten to twenty minutes. Finally the mass is brought to a boiling point ; strain and serve. Gelatine may be added and the mixture served cold as a jelly. Cooked tomatoes, onions, celery, or other flavoring suited to individual tastes may be added at the beginning of artificial digestion.

BEEF TEA.

Free a pound of lean beef from fat, tendon, cartilage, bone, and vessels ; chop up fine, put into a pint of cold water for two hours. Simmer on the stove three hours, but do not boil. Make up for the water lost by adding cold water so that a pint of beef tea represents one pound of beef. Press the beef very carefully, and strain.

BEEF JUICE.

Cut a thin, juicy steak into pieces about one and one-half inches square. Sear separately one and one-half minutes, on each side, over a hot fire. Squeeze in a hot lemon squeezer, flavor with salt and pepper. May add to milk, or pour on toast.

MEAT CURE.

Procure slices of steak from the top of the round
without fat.   Cut meat into strips, removing all fat,
gristle, etc., with a knife.   Put meat through mincer
at least twice.   Then beat it well in a roomy sauce
pan with cold water or skimmed beef tea, to the
consistency of cream.   The right proportion is one
teaspoonful of liquid to eight of pulp.   Add black
pepper and salt to taste.   Stir the mince briskly with
a wooden spoon the whole time it is cooking, over a
slow fire, or on the cool part of cupboard range, until
hot through and through and the red color disap-
pears.   This   requires   one   and   one-half   hours.
When done it should be a soft, stiff, smooth puree,
of the consistency of good paste.   Serve hot.   Add
for the first few meals a softly poached white of an
egg.

OATMEAL GRUEL.

One half a cup of coarse oatmeal, three cups boiling water, one teaspoonful of salt, and cream.   Add
oatmeal and salt to boiling water, and cook three
hours in a double boiler.   Force through a strainer,
dilute with cream, reheat and strain a second time.
Serve with salt or sugar.

CREAMED CHICKEN.

One tablespoonful of butter and one of flour, and
add to that half a pint of cream, a little salt, pepper,
and celery salt and the meat from half a chicken
which has been put through the meat grinder.

CREAMED EGGS.

⅓ glassful of chicken
stock.
⅓ glassful of cream.

4 eggs.
½ teaspoonful of salt.
Pepper to taste.

Heat together the cream and the stock in a double boiler. Beat the eggs without separating, and stir into it slowly. Stir until thick, season and serve. This is the most nourishing preparation of eggs for an invalid.

CREAMED CALF BRAINS.

Parboil the brains. Blanch them and cut into small pieces. Put into a double boiler one tablespoonful of butter and a scant one of flour. Add half a pint of cream. Put in slowly the beaten yelk of one egg, stirring constantly. Season with salt and pepper, add the brains, cook three minutes, and serve on toast.

CREAMED OATMEAL.

Boil oatmeal as for breakfast, rub it through a fine sieve, add a little cream, and cook very slowly in a double boiler for half an hour longer. When perfectly smooth, add a very little salt and rich cream.
This the most delicate preparation of oatmeal that an invalid can take.

CREAMED SWEETBREADS.

Make sauce as for creamed chicken. Add parboiled sweetbreads chopped very fine and a tablespoonful of sherry wine.

6

APPLE SOUP.

Two cups of raw apple, two cups of water, two teaspoonsful of corn starch, one and a half tablespoonsful of sugar, one saltspoonful of cinnamon, and a bit of salt. Stew the apple in the water until it is very soft. Then mix together in a smooth paste the corn starch, sugar, salt, and cinnamon with a little cold water. Pour this into the apple and boil five minutes. Strain it and keep hot until ready to serve. May serve with cream if desired.

BEEF MINCE.

Have a pound of beef from the round. Free it from all sinews and fat. Mince it very fine. To two tablespoonsful of butter in a sauce pan put in the meat and a teaspoonful of onion juice. Stir for three or four minutes, or until the meat is hot through. Add salt and pepper, and if desired a little lemon juice. Serve on hot buttered toast.

PANNED OYSTERS.

Put two tablespoonsful of butter in a saute pan. Lay twenty good sized oysters into it. When the edges curl and the oysters plump, dust them with pepper and salt, and serve at once on toast. Two tablespoonsful of sherry can be added before serving if desired.

RAW MEAT DIET.

Scrape pulp from a good steak, season to taste. Spread on slices of bread, then sear the bread slightly and serve as a sandwich.

FLAXSEED LEMONADE.

One tablespoonful of whole flaxseed, one pint of boiling water, lemon juice, and sugar. Pick over and wash the flaxseed, add water, and cook two hours, keeping just below the boiling point. Strain ; add lemon juice and sugar to taste.

ORANGE ADE.

Juice of 1 orange.   1 ½ tablespoonsful of syrup.
2 tablespoonsful of crushed ice.

Make a syrup by boiling eight minutes one cup of water and half a cup of sugar. Mix the orange juice and the syrup, and pour over the crushed ice.

SHERRY NOGG.

To the yelk of one egg thoroughly beaten add one tablespoonful of powdered sugar and two tablespoonsful of sherry wine and a pint of whipped cream.

DAINTY MENUS

FOR CONVALESCENT PATIENTS.

———

Select the daintest of tray covers and china, and make the tray look as attractive as possible in every way.

No. 1.

Bouillon.
Creamed Chicken on Toast,
garnished with parsley.
Bread and Butter Sandwiches, served on lettuce leaf.
Small Mold Bavarian Cream
with whipped cream.

No. 2.

Cream of Celery Soup.
Supreme of Chicken with White Sauce, garnished with
parsley.
Beaten Biscuit.
One Fresh Tomato, garnished with chopped celery or
Nasturtium leaves.
Mold of Wine Jelly.

No. 3.

Broiled Breast of Chicken with drawn butter.
Creamed Sweetbreads on Toast with peas.
Bread and Butter Sandwiches.
Cup of Delicate Chocolate.
A Little Whipped Cream, frozen.

No. 4.

An Orange cut in half, after being on ice several hours.
Broiled Sweetbread, garnished.
Quail on Toast.
Celery Salad, garnished with celery tops.
Bread Sticks.
Pineapple Ice.

No. 5.

Oyster Soup.
Fish Coquille in a nest of watercress or parsley.
Broiled Beef Tenderloin, mushroom sauce.
Parisienne Potatoes.
Light Rolls.
Brandy Peaches.

No. 6.

Sweetbread Croquettes with creamed peas.
Bread and Butter Sandwiches.
Celery Salad.
Chocolate with whipped cream.
Plain Ice Cream.

CHAPTER ON MENUS.

INFORMAL DINNER.

No. 1.

Salted Almonds. Olives.
Chicken Gumbo.
Fish Pudding. Parisienne Potatoes.
Roast Turkey, Cranberry Sauce. Croquettes of Peas,
Asparagus, and Stuffed Sweet Potatoes.
Celery Salad.
Charlotte Russe, or Ices.
Coffee.

No. 2.

Salted Almonds, Pickles, and Celery.
St. Germain Soup.
Broiled Pompano. Potatoes au Gratin.
Grouse or Pheasant. Asparagus. Peas.
Nut and Celery Salad.
Fig Pudding with fancy sauce.

DINNER.

No. 1.

Salted Almonds.   Maron Glace.
Blue Points on Shell.
Consomme.
Lobster Timbals with lobster sauce.
Fillet of Beef.   Parisienne Potatoes.   Asparagus.
Victoria Punch.
Stuffed Quail.   Croquettes of Peas with white sauce.
Stuffed Mushrooms.
Celery Salad.
Fancy Ices and Cakes.
Coffee.

No. 2.

Blue Points on the Shell.
Consomme.
Stuffed Lobster.
Fillet of Beef.   Creamed Cauliflower.   Potatoes.
Roman Punch.
Broiled Grouse with asparagus.   Sweetbread Croquettes
with peas.
Green Grape Salad.
Fancy Ices and Cakes.
Coffee.

No. 3.

Oyster Cocktail.
Cream of Celery Soup with whipped cream garnish.
Lobster a la Newburg.
Venison Steaks.   Asparagus.
Roman Punch.
Sweetbread a la Victoria, Allemande Sauce.   Peas.
Salad a la Jardin, in Turnips.
Sultana Roll Ices, with claret sauce.   Cakes.
Coffee.   Cheese.   Crackers.

No. 4.

Cavierre on Toast.
Consomme.
Lobster Timbals.
Fillet of Beef.   Stuffed Sweet Potatoes.   Asparagus.
Fruit Punch.
Pheasant.   Potatoes en Surprise with sauce.
Stuffed Mushrooms.
Waldorf Salad.
Fancy Ices and Cake.
Coffee.

SIMPLE LUNCHEON.

No. 1.

Sliced Pineapple with crushed ice and sherry.
Bouillon.
Oyster Patties.
Stuffed Lamb Chops with peas.
Egg Salad.
Brick Cream and Cakes.
Coffee.

No. 2.

Puree of Asparagus with whipped cream garnish.
Oyster en Coquille.
Chicken Croquettes with creamed peas.
Celery Salad.
Bavarian Cream.   Macaroons.
Coffee.

No. 3.

Tomato Puree.
Mushrooms a l'Algonquin on Toast.
Broiled Fillets.   Potatoes en Surprise.
Hollandaise Punch.
Pepper Timbals.   Chicken Salad.
Individual Orange Ice with Cakes.
Coffee.

No. 4.

Grape Fruit.
Bouillon.
Fish Croquettes with white sauce.   Potatoes.
Broiled Quail on Toast with asparagus.
Hollandaise Punch.
Little Pigs in Blankets (Sweetbreads).
Waldorf Salad.
Individual Ices and Cakes.
Coffee.

No. 5.

Large Pink Grapes.
Crushed Ice with sherry wine.
Lobster Cutlets with Bechamel sauce.
Broiled Grouse.   Potatoes en Surprise with oyster sauce.
Victoria Punch.
Croquettes of French Peas with sauce.
Salad a la Jardin.
Individual Brick.   Cake.
Coffee.

No. 6.

Oyster Bisque.
Fish Croquettes with Potatoes.
Broiled Quail.   Saratoga Chips.   Asparagus.   `
Punch.
Supreme of Chicken with Bechamel sauce.
Green Grape Salad.
Ice Cream Plates with Brandied Fruit.   Cakes.
Coffee.

MEMORANDA.

The following blank pages are intended for recording receipts that may hereafter come to notice, or for any data on those contained herein.